CHALMETTE NATIONAL HISTORICAL PARK

THE FAMOUS BATTLE of New Orleans was fought here. This 1840 plantation home facing the river is the René Beauregard House, owned after 1880 by the son of General Beauregard of the Confederacy. The monument at the Battlefield is a hundred foot marble obelisk.

As it has been throughout the recorded history of the Mississippi River, any country that had designs on possessing the United States, knew that they had to control most of the river. In the war of 1812, the British hoped to regain their American Colony and in the final year of the war, they attempted to capture New Orleans. They contacted Jean Lafitte, the so-called pirate patriot, who pretended to agree to help them slip through back waterways to New Orleans, but he warned Governor Claiborne of their plans.

On December 23, 1814, the British vanguard advanced as far as Chalmette and set up camp. General Andrew Jackson's men marched 120 miles in two days, through bayous and cypress swamps, to stop them. The British referred to Andy's men as "Kaintucks" and "dirty shirts", but these backwoodsmen with their long rifles, accurate marksmanship, tomahawks and hunting knives, gave the British a bad time. The battle was fought on January 8, 1815, with neither side aware that on December 24th peace had already been declared in the Treaty of Ghent. Fearing that the city would be burned, the citizens fought desperately and won. General Andy Jackson has always been a hero to the city. The French people of New Orleans loved his free spirit and erected a monument to him.

Page 1

JACKSON SQUARE

THE HEART OF NEW ORLEANS

Jackson Square is the heart of the French Quarter (Vieux Carré or Old Square) in New Orleans. The St. Louis Cathedral, stands tall at the back of it, flanked on either side by the Cabildo and the Presbytere, once official buildings of the Spanish Government. Andy Jackson proudly rides his horse in the middle of the walkways, and the square is completed on the other two sides by the famous Pontalba apartment buildings of 1849, built of red brick with lacy wrought iron balconies on the upper levels. At the front of the Square, visitors can visit the shops, ride in carriages, or climb the stairs to the Moonwalk where sight-seers and lovers wander and gaze at the river traffic.

The city of New Orleans grew up around the French Quarter, stretching uptown where the Americans settled, and downtown into the Faubourgs (Suburbs) where the aristocratic Creole French preferred to live. Its streets were built parallel to the river in a series of semicircles, which gives it the nickname of the Crescent City, for it has the shape of a half moon. There are tall downtown buildings just on the other side of Canal Street (which used to be a canal).

Across the river is Algiers, which is really a part of the city. In spite of its close association with New Orleans, it has always maintained its own distinct character. Some believe its name came about because of its being linked with the activities of the pirate Jean Lafitte, thus inviting comparison with the country of Algiers which was formerly a notorious pirate territory.

PADDLEWHEELERS AND EXCURSION BOATS

New Orleans is like an island surrounded by water. It lies between the Mississippi River and Lake Pontchartrain. Since it cannot grow in width, it grows in length upriver and downriver. It has spread out of Orleans Parish into three adjoining parishes (counties), one of which is across the Mississippi. The two sides of the River are connected by bridges, built as fantastic engineering feats, because there is no bedrock to build on.

The city is protected by a series of connected levees (earthen dikes) around the entire area. This is an essential precaution for the city's protection, since it is low and flat, and the part around Audubon Park is actually below sea level. A system of pumping stations was installed many years ago to pump off all the surplus rain water that falls on the city. When the river is at flood stage, one can stand on the levee and see the level of the river raised higher than the streets on the other side. At times one can see ships standing so high on the riverside that the entire superstructures are visible to people on the land side, and they almost look as if they were perched on top of the levee.

Obviously, both the natives and the visitors to the city are fascinated with boats and ships, particularly the steamboats that take passengers out morning, noon, and night. Some of the familiar names of the well-loved New Orleans paddle-wheelers are the Creole Queen, Steamboat Natchez, Cajun Queen, Cotton Blossom, Mark Twain, Jean Lafitte, and Voyageur.

Page 4

RIVER TRAFFIC

New Orleans is one of the busiest ports in the United States, even though the ships, with the help of special river pilots, must make a hazardous and long journey from the Gulf of Mexico through a pass at the mouth of the river which is most difficult to navigate, and upstream through treacherous and powerful currents and sand bars.

These pilots, and all of the pilots that work other stretches of the river, must have an intimate knowledge of each sand bar, and of how it shifts with time, as it is not possible for all of the captains of foreign and domestic ships to acquire this knowledge.

A short distance upriver from New Orleans, at the point of the shortest route from the Mississippi River to Lake Pontchartrain, the United States Corps of Engineers constructed the Bonnet Carré Spillway. The spillway dam, constructed at the bank of the river, has nearly twice the flowage capacity of Niagara Falls. 250,000 cubic feet of water per second can be diverted into immense Lake Pontchartrain, through a dike enclosed runway, to relieve the pressure on the levees when the river is at flood stage.

The Superdome, which has come to dominate the aerial view of the city, also has problems with rain. Rainwater which falls on the vast area of the dome must be collected in tanks, and then released gradually, or it would flood the area. Often when heavy rains occur, the pumping stations cannot keep up with the amount of water that is falling on the city until the rain has ceased.

Page 7

THE NEW CITY AND THE OLD QUARTER

The juxtaposition of old and new "Nouvelle Orleans" is startling. The tall hotels and commercial buildings on the upriver side of Canal Street are so close to the small old homes, shops and inns of the Vieux Carré, that they almost cast a shadow across it. One walks from a totally modern world of metal and glass facades that mirror the trees and the blue sky, into an old world left over from Europe, influenced by the Spanish and French people of the past. The lace balconies make one walk along the sidewalks looking upward. The shops are on the first floor, and behind them, the original owners had a private courtyard, with the family quarters on one or two floors above the shop. Carriages still transport visitors around the narrow streets, and it is possible to lunch in cool patios festooned with vines and even balloons.

AN ISLAND SURROUNDED BY WATER

Outside the city of New Orleans, East and West of the dry land, lie enormous areas of swampland. Other sizable rivers and small lazy bayous intersect every area of land throughout Louisiana. Shrimp boats dock at many settlements, and Cajun cabins, with steps leading to the loft where the boys slept, are still to be seen. Alligators lie in the green scum of the wetlands, seemingly in a stupor, but their moving eyes are always searching for a possible meal. Boat trips take visitors to see huge cypress and tupelo gum trees with their swollen bases anchored in the water. This is "Cajun" Country. The word is a corruption of Acadian, from the people driven out of Canada.

DESTREHAN

PLANTATION COUNTRY IN LOUISIANA

It is interesting to read what early travelers' impressions were about the plantation country in the lower river area.

From the famous notes of Mrs. Trollope (who was not always happy with America): *"The unbroken flatness of the banks of the Mississippi continued unvaried for many miles above New Orleans; but the graceful and luxurient palmetto, the dark and noble ilex (holly), and the bright orange (many plantations had orange groves then), were everywhere to be seen, and it was many days before we were weary of looking at them."*

In Captain Basil Hall's notes: *"The district of country which lies adjacent to the Mississippi, in the lower parts of Louisiana, is everywhere thickly peopled by sugar planters, whose showy houses, gay piazzas, trig gardens and numerous slave villages, all clean and neat, gave an exceedingly thriving air to the river scenery."*

Destrehan Plantation, near the town of the same name, was an early plantation house, completed in 1790. Many of the very early Louisiana planters' homes showed the influence of the West Indies, where houses were built to keep their occupants cool. The main floor was raised, and the ground floor was used as a basement. The kitchen, wine cellars, and often a summer dining room were some of the rooms to be found on this floor. The formal dining room, bedrooms, and parlors were located on the floor above. The roof was high and it extended far out over a wide gallery to provide shade. Windows were used as doorways to permit the summer breezes to keep the house cool. The window sashes could be raised higher than one's head, and hot rays of sunshine could be shut out by closing the slatted shutters (called jalousies).

The ceilings were not as high in these early homes as in some of the later, grander homes that were built when the planters became extremely wealthy. The planters began to bring in artisans from Europe to make woodcarvings, and ornate plaster trims for their 15 to 20 foot ceilings, such as cornices and center medallions.

Then they began to paint the wooden doors, their frames, and the baseboards with decorative "faux bois" (false wood grain) and "faux marbe", (false marbleizing), which was also used on the mantle pieces.

The homes were built of cypress, which termites did not attack, but the floors were usually of heart pine, because cypress was too soft.

CROPS OF THE PLANTERS

Cotton is probably the crop most frequently associated with the Mississippi River. The cotton planters that became millionaires in a few short years built many fabulous plantation houses, some still standing. They came to the Mississippi in the early 1800's to seek their fortunes and they made them by sowing cotton seeds in black alluvial soil.

Writers often speak of the rich soil of the "Delta" in these regions and we hear a few speak more technically of "deltaic soil". The true delta, which is presently and continually forming, is at the mouth of the Mississippi where the river's heavy load of silt is continually being dropped, forming new land. In the larger perspective, the old delta where the rivers diverge and where various former paths of the Mississippi lie, is a large triangle (which delta means), marked roughly on its western boundary by a diagonal line drawn from Baton Rouge through Lafayette, Louisiana. The diverging rivers in this area, such as the Atchafalaya, are called distributaries, in contrast to the tributaries that flow into the Mississippi farther upstream.

In southern Louisiana, sugar cane has long been an important crop. Along with cotton, indigo was a very important crop in very early years. Near the end of the 19th Century, an epidemic of caterpillars, which the French called "chenilles", was destroying the cotton and indigo crops. Conveniently, at about this time, Étienné de Bore in New Orleans perfected the process of crystallizing sugar so that it could be stored or shipped. Most of the planters then began to grow sugar cane. The cane is cut and hauled from the field to a cane loader by tractor. There it is loaded on enormous cane trucks and sent to the sugar house where the juice is extracted from the crushed cane, then clarified, boiled, and crystallized. Harvesting continues through 12 hour days, seven days a week, from the middle of October through December. Recently, soy beans have been replacing some of the cane fields, and many plantations have always grown some rice, which they irrigate from the Mississippi. Farther up on the river, winter wheat has now become one of the "southern" crops. It is grown as far south as St. Francisville. Wheat fields waving in a breeze present an undulating sea of green that is a beautiful sight indeed.

Other new "agricultural crops" include crawfish and catfish farming, and the products are shipped out to a worldwide market. (Note that the word "crayfish" is practically unheard of in the area).

Page 13

GOTHIC STEAMBOAT ARCHITECTURE

IN ANTEBELLUM DAYS, before the "War Between the States", as Southerners often preferred to call the Civil War, the paddlewheel boats really came into their own. Robert Fulton invented the steamboat, but he surely did not realize the grand form it would take, with all of the ornate gothic trim decorating it. An old timer described it as "an engine with a raft with $11,000 worth of jig-saw around it," and he wasn't far wrong.

Mark Twain described some of the elegance that passengers found on the fine big steamboats. "The pilot house, hurricane deck and boiler deck guards were all garnished with white wooden filigree work of fanciful patterns. All along the ceiling of the snow white inner "cabin" were curving patterns of filigree-work touched up with gilding, stretched overhead, with big chandeliers every little way, each an April shower of glittering glass drops. The colored glazing of the overhead skylights cast a rainbow light below. It was a long resplendent tunnel."

How natural it was then, for some of the "river men" to decide that they would like to apply that particular architecture to the homes they built.

San Francisco Plantation is one of the few remaining examples of this unique architecture. It was a fanciful house, built by a Frenchman with a fanciful name, Valsin Marmillion. He spent such a quantity of money achieving his dream that he nicknamed it with a phrase, "Sans Frusquin", that roughly stood for our "without a red cent" description. The natives in the area corrupted the name to San Francisco. San Francisco was used as the title and for the picture on the dust jacket of the book "Steamboat Gothic" by Frances Parkinson Keyes.

SAN FRANCISCO

Page 14

OAK ALLEY

AN ALLEÉ OF OAKS

The original elegant French name of this plantation house was "Bon Séjour" (Pleasant sojourn). It is a large, but delicately designed home, and was built by Jacques Telesphore Roman, a brother of a former governor of Louisiana. (Note the incorporation of a Greek name: "Telesphore". Names from Mythology were often used by both Creole and Acadian French.)

All the ship captains and the passengers that passed Vacherie on the packet boats began to refer to the magnificent alley of oaks that lead from the levee to the entrance as a landmark, and so referred to it as "Oak Alley", and that is the way we all know it now.

Page 16

It is an enormous and impressive home, almost symmetrical on all four sides, with galleries all around. The architect incorporated all of the elements that make a home in the hot humid South liveable.

Visitors from the North are always surprised when we point out the Live Oak trees. The leaves are very small, and they think these trees must not be true oaks. Both the Live Oak (Quercus virginiana) and the Water Oak (Quercus nigra) are very much a part of the family. However, they both have small leathery leaves (probably to cut down on water loss) and they are both evergreen. They live many years and there are records of them being at least 300 years old. As they get older, their limbs begin to weigh heavily and therefore bend and grow horizontally until they touch the ground or have to be supported. These huge limbs become great playgrounds for children, who can easily clamber up on them from the ground.

TEZCUCO

AN ANTEBELLUM "COTTAGE"

This house with the Aztec name, was built by a grandson of Marius Pons Bringier, a very early French planter who had large holdings in the area of Convent and Burnside on the West Bank of the Mississippi in Louisiana. His children and grandchildren built almost all of the manor homes in the area with the exception of Houmas House. They included Ashland (now Belle Helene), The Hermitage, Bocage, Union, Bagatelle, Colomb House, and the Zenon Trudeau House, some now gone.

Benjamin Tureaud was the builder of Tezcuco. He was the son of Marius' daughter Betzy who at 14 married an old Judge, Augustin Tureaud (no relation to Trudeau who another family member married), and they built Union Plantation. Benjamin built on adjoining land, and in its day, Tezcuco was called a "raised cottage", although the ceilings are 15 feet high and some of the rooms are 25 feet square. Benjamin had fought in the Mexican wars, and had apparently brought the name from Mexico. Tezcuco was the place where Mexico was founded and where Mexico City is now, on Lake Tezcuco, a place of beauty with floating gardens. The Aztec spelling has been changed to the Mexican form, Texcoco — now a dry lake bed.

Tezcuco kept the plantation architecture with large roof, galleries all about, and tall shuttered windows, but the iron filagree was a new touch that began to be adopted by builders of plantation houses.

ROSEWOOD

A GRAND RECREATION

Travelers often look at the famous plantation homes in Louisiana and Mississippi and wonder how people ever managed to build them, what they looked like when they were new, how the trees got there, and so on.

James Ellis Richardson has spent more than twenty years building the re-creation of a mid-nineteenth century Louisiana plantation home. This is not a restoration. There was only bare ground when he began his project. It is a 25,000 square foot, air-conditioned structure that includes both structural and architectural elements gathered from almost forty Louisiana plantation homes. These houses had been demolished, sadly, for various reasons. His home thus provides a living link with the rich cultural history of the region.

The home is called Rosewood and is located in Brittany, Louisiana, 3 miles from Sorrento and about 8 miles from the Mississippi on Highway 431. The landscaping is continuing and the trees are being planted. The whole phenomenon of establishing a home and its environment is still in process. At the same time that he has been building the house, James Richardson has been traveling and amassing an extensive collection of Victorian Rococo furniture made of rosewood, mahogany and walnut, precious porcelain and crystal objects d'art, and artifacts to furnish the large kitchen that make it seem as if it has always been there.

Original documents of the specifications needed for building many of the old homes are on record. If you would take one of these documents to a contractor today, he would probably laugh you out of his office. They specify wood without any knotholes whatsoever, of cypress for the construction and heart pine for the floors. The plaster is to be allowed to slake (soak in water) for a long time before stirring (to make it harder), and marble dust is to be added to the plaster for the ceiling. The ingredients for the paint are specified differently for each of three coats of paint, and the finest components of all are saved for the final coat.

Page 19

BOCAGE

HOMES OF THE PLANTERS

Early travelers on the Mississippi were always impressed by the stretch of river between Baton Rouge and New Orleans, where the land was occupied by sugar plantation.

Mark Twain's description: "From Baton Rouge to New Orleans, the great sugar plantations border both sides of the river all the way, and stretch their league-wide levels back to the dim forest walls of bearded cypress in the rear. Shores lonely no longer. Plenty of dwellings all the way, on both banks — standing so close together, for long distances, that the broad river lying between the two rows becomes a sort of spacious street ..... And now and then you see a pillared and porticoed great manor house embowered in trees."

Houmas House at Burnside is probably one of the purest and finest examples of architecture in the United States—a true classic. The eight sided garconnieres (houses for the young boys to inhabit), one on either side at the rear, are unique.

Nearby, Bocage was built by another Bringier daughter and her husband, Christophe Colomb, said to be a French relative of Christopher Columbus. The replica of an old "powder house" holds a water cistern.

Hermitage was built by Michel Doradou Bringier, and is a very early home with smaller rooms and lower ceilings, but with columns on four sides. A reconstructed peak roofed well-house decorates the garden.

SHADOWS ON THE TECHE

FOLLOWING THE BAYOUS FROM THE MISSISSIPPI

In the "good old days" travelers could take a side trip down Bayou Manchac to the East, or down Bayou Lafourche to the West and connect with the Atchafalaya River, a former path of the Mississippi, and that brought them to "Acadiana", a region that is the land of the Cajuns.

Shadows on the Teche was built on the bank of Bayou Teche in New Iberia, a lazy wide stream in the heart of Cajun Country. However, the builder was not a Cajun, but a man of English descent, David Weeks. It is a tastefully designed home, built of rosy pink bricks. The handsome white columns across the front and the dormer windows are typical of the period, but the unique accent is provided by four sets of protective green jalousies that shelter each end of the upper and lower galleries.

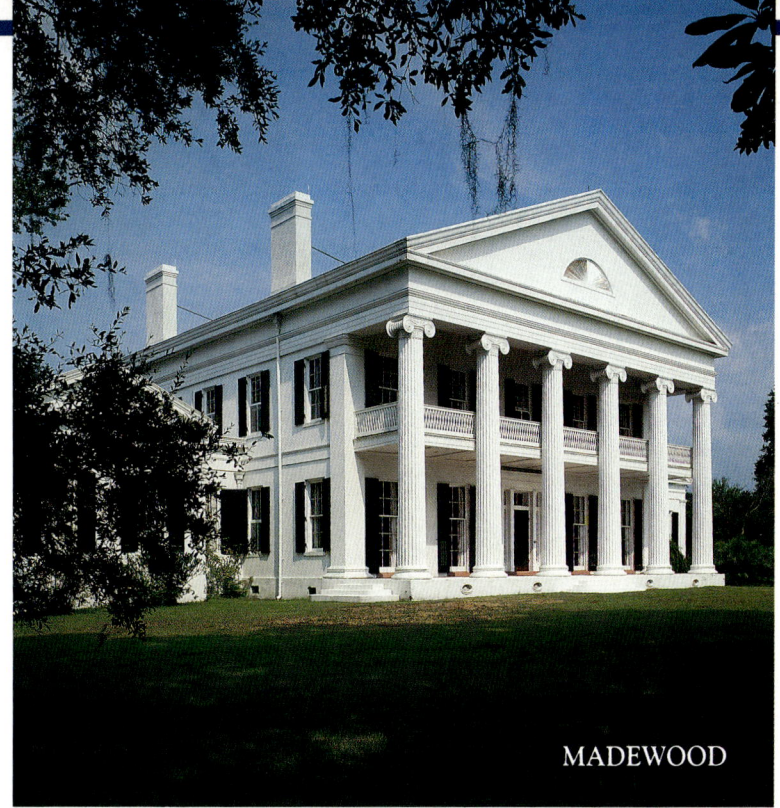

Madewood, on the banks of Bayou Lafourche in Napoleonville, is a classic house with ionic columns and wings at either side. It is massive in scale, but simple and pure in style. A large ballroom was incorporated into the house.

MADEWOOD

Mintmere Plantation is in New Iberia. It is a Greek Revival raised cottage, built in the mid-nineteenth century, and was used by General Alfred Lee as headquarters when Yankee troops occupied the town in 1863. On the same grounds is the 1790 Armand Broussard House, an intact mud and moss plantation structure of great architectural and historical significance.

MINTMERE PLANTATION

Longfellow-Evangeline State Park, also on Bayou Teche, was developed around an 1850 Acadian House that was originally believed to have belonged to Gabriel, Evangeline's lost lover, although the story is probably apocryphal. The house is, nevertheless, an outstanding example of that type of architecture and has a typical kitchen-garden with okra, hot red peppers, eggplant, etc.

ACADIAN HOUSE

GARDENS IN THE MARSHLANDS

West of the Mississippi in Louisiana, the land dries out and salty and sawtoothed grasses appear instead of the dense vegetation of the swamplands. They create marshlands, the so-called wet prairies of Louisiana. Distributaries from the Mississippi run through it, and

JUNGLE GARDENS

JOE JEFFERSON HOUSE

JAPANESE TEA HOUSE

LIVE OAK GARDENS

occasionally one sees "islands of high ground" rising from the marshes. These are created by the pushing up of underground domes of salt, which can be mined. Homes and gardens were built on these domes, which were often called "chenieres", i.e. oak islands, from the French word "chêne" for oak. They were built as if they were reaching for the sky and the fresh air, and getting their feet out of the wet marshes.

The wonderful house at Live Oak Gardens at Jefferson Island near Delcambre, Louisiana was built by the famous actor, Joe Jefferson, noted for his portrayal of Rip Van Winkle. There are porches and gables and gingerbread projecting in every direction, just inviting the visitor to come and see what is inside. The gardens themselves were established by J. Lyle Bayliss who came down from Kentucky to live on beautiful Lake Peigneur. In the gardens, is an ancient and gigantic oak, where President Cleveland used to relax and read in the shade of its great arms, and so it was named for him. A tea house, of authentic design was added to the gardens, shaded by a cypress tree and reflected in a pool with stepping stones.

Nearby on another salt dome, at Avery Island, are the famous Jungle Gardens, created by Walter McIlhenny. An ancient Chinese Buddha was brought from the Orient and placed in a blue tiled pagoda on a hill at the end of a lake. Both of these gardens are far enough south to have subtropical vegetation.

NOTTOWAY, THE PINNACLE OF AN ERA

Nottoway was built for one of "The Randolph's of Virginia", John Hampden Randolph, by the famous architect, Henry Howard. He wanted a grand and palatial home, different from anything that had been built in Louisiana, and he succeeded in getting just that. It is upriver from the small town of White Castle on the west bank. The town did not receive its name from Nottoway, even though it looks like a "white castle", but from another house that was divided and moved.

Nottoway stands today as the largest existing antebellum plantation home. It is not only wide and very tall, but the dependencies attached to the back make it great in length. The kitchen garden is usually planted in red tulips in Spring, and the red salvia in the Autumn attracts hummingbirds by the dozens.

The ballroom became the most famous room because everything in it was white, even the floor; and the woodwork and decoration were especially ornate and beautiful. The curved wing at the right side of the house is the exterior facade of the ballroom.

The bricked courtyard in the rear has a tropical appearance, with banana trees and leathery dark greens.

MAGNOLIA MOUND

RIVER CITY CAPITAL

Baton Rouge is located on the River and has long been the Capital of Louisiana. The State Capitol building was commissioned by Huey Long. He admired the Capitol Building in Nebraska, but wanted his to be the tallest, so that is the way it was built. It is surrounded by gardens laid around old live oak trees, and blooming camellias and azaleas.

There are not too many old buildings in the city, because a large part of the original "Spanish Town" was burned by the Union army.

However, there are several old homes that were originally a part of plantations and two of these have become tourist attractions. Magnolia Mound is so named because it was built on an Indian Mound and is situated in a grove of magnolia trees. It dates from the late 1700s and has been restored by the City of Baton Rouge and the Foundation for Historical Louisiana.

Mount Hope was built in 1817 by a German planter who never meant it to be an elaborate showcase of a house. It is pure of line and devoid of frill and pretense, but possesses all the characteristics of a typical, and very practical, early southern plantation home.

It also has great live oak trees in the front and back, and there are two replicas of outbuildings on either side of the front yard. To the left is a latticed and curtained gazebo, and to the right is a very large brick pigeonnier (dovecote).

MOUNT HOPE

A RED STICK AND A MEDIEVAL CASTLE

Shortly after Sieur d'Iberville landed on the Mississippi Coast in 1699, he led an expedition up the Mississippi River as far as Louisiana's Red River. About five leagues above Bayou Manchac, he wrote about finding high bluffs on which stood a reddened post which the Indians called "Istrouma". In his French language, this meant "Baton Rouge" (Red Stick). It had been placed there to mark the division of land between two Indian nations. It is now memorialized by a piece of sculpture placed on the River Front Plaza.

The U.S.S. Kidd is a destroyer that is permanently docked in Baton Rouge as a tourist attraction. It is painted in camouflage colors. A Maritime Museum has been built in connection with it. Many veterans of World War II who sailed on destroyers have reunions at this facility.

The Louisiana Arts and Science Center, built in the old railroad depot, is a children's museum, and there are old railroad cars on exhibit there that the children can visit.

Baton Rougeans have a great affection for the Old State capitol, but Mark Twain scoffed at it. He believed that Sir Walter Scott was partly responsible for it, because he said, *"It is not conceivable that this little sham castle would ever have*

been built if he had not run the people mad ... with his medieval romances."

The Old State Capitol, U.S.S. Kidd, the museums and plaza are all in close proximity so that visitors walk along the levee and browse through the buildings at their leisure.

ROSEDOWN PLANTATION

GARDENS AND GAZEBOS

Rosedown, surrounded by landscaped gardens, fountains, gazebos and massive live oak trees, is one of the most visited homes in Louisiana. It was built by Daniel Turnbull, a wealthy cotton planter in St. Francisville. The Turnbulls shipped their cotton to New Orleans by riverboat and had marvelous wall coverings, exquisite New Orleans furniture, chandeliers, silver, and marble statuary sent back up river to them. Martha Barrow Turnbull, Daniel's wife, was a remarkable horticulturist. The imported camellias, azaleas and the sacred cedar from Japan attest to her dedication.

Page 33

OAKLEY HOUSE

AUDUBON'S HAPPY LAND

Just above St. Francisville, the Mississippi River stops being the boundary between the states of Mississippi and Louisiana as it rushes down from the North, and just cuts right on down past the "Florida Parishes" of Louisiana and continues toward New Orleans and the Gulf. Louisiana has parishes rather than counties, and these parishes east of the Mississippi, in the "toe" of Louisiana's "boot", were once part of West Florida under Spanish rule.

In this gracious community, John James Audubon found a niche appropriate to his talents. He was hired at Oakley by Mrs. Eliza Pirrie to teach art to her daughter, for which he was paid $60.00 a month plus room and board. He also taught dancing to the young people of the area and in his spare time, he was able to paint birds. His situation was so satisfactory that he sent for his wife, who established a private school of her own. Audubon painted his Wild Turkey here, the most coveted of all his prints today.

Oakley House is situated in Audubon State Commemorative Area, just south of St. Francisville. This is one of the most satisfying places in Louisiana to just take a walk in the woods. It is easy to see why Audubon called this his "Happy Land".

Page 34

ASPHODEL

To the east of St. Francisville, near Jackson, Louisiana, Asphodel hugs the crest of a hill, surrounded by azaleas and trailing ivy. Its builder apparently had a great love for the classics. He chose the early Doric style of classical Greek architecture and picked the name Asphodel (referring to daffodils and narcissus) from classic literature. The two wings are the most unique part of the house.

There is a sort of hidden abandoned garden down the hill to the left of the house and it is reached by a set of worn old brick steps. The azaleas here, growing in old leaf mold, are lush in the Springtime.

THE COTTAGE

ENGLISH LOUISIANA

St. Francisville is the heart of the Florida Parishes in the toe of Louisiana, and it was the settlement to which the English migrated from the East to the banks of the Mississippi River. Most of the family names that still persist here are English.

The Cottage (above) was one of the early homes, and it was built on a Spanish land grant in 1795, because the Spanish were in control of the area around Baton Rouge for a time. It stands on an old fringed bluff in a park-like setting, and the long front porch, complete with rocking chairs, invites one to relax in the cool breezes and watch the Spanish moss move gently. The house was periodically added to by Judge Thomas Butler between 1811 and 1859, until the present combination was produced. The additions were so well planned that the pleasing structure seems to have been designed all at the same time.

The Myrtles (below) is an 1830 house built by Judge Clark Woodruff. He commissioned intricately detailed woodwork, center medallions for the ceilings, friezes, silver doorknobs and handsome marble mantlepieces for his home. The front gallery runs the length of the house and wraps around the sides, and all of it is framed with railings and supports of lacy cast iron entwined with a design of grapevines.

Afton Villa was a brick Victorian-Gothic type of house that almost looked like a castle with its sharply peaked roofs. It was very English in appearance. Tragically the house burned, but the beautiful gardens remain (top right). The long road leading into the gardens is framed by live oak trees and bordered by large old azalea bushes. Near the entrance a lovely fountain cools the atmosphere.

THE MYRTLES

Page 36

AFTON VILLA GARDENS

PARLANGE

BY AN OXBOW LAKE

Just a ferry boat ride east across the Mississippi brings us into the French world of Louisiana again. Parlange is near New Roads on a lake made by a loop of the river that was cut away by the main stream many years ago. Parlange was built by the Marquis de Ternant and furnished with exquisite French antiques. His widow married Colonel Charles Parlange and the house has remained in the family ever since. It is an early raised house, complete with twin pigeonniers.

Page 37

ROSEMONT

A STATESMAN'S BOYHOOD HOME

Jefferson Davis was such a hero to the Confederacy that every place he ever lived has become a sort of shrine. He spent his last years at his beloved Beauvoir on the Gulf Coast of Mississippi, but Rosemont, near Woodville, was his boyhood home. This is just north of St. Francisville where the Mississippi River becomes the dividing line between Louisiana and Mississippi. You can imagine him as a boy, walking the paths and climbing the split rail fences that surround this charming house, or playing on the back gallery that overlooks a wooded ravine. It is a simple, but charming cottage, furnished in the taste of a comfortable country family. In the Springtime,

Page 38

white dogwood blossoms and the pale pink of wild azaleas (called honeysuckle by the natives) make the large yard a fairy land.

OVERLAND TO THE MISSISSIPPI

Natchez was the connection between the Mississippi and the rest of America. Travelers came down the Natchez Trace from Nashville, Tennessee to get a boat headed for New Orleans. This was also a place where one could cross the river and connect with the El Camino Real which went through Spanish territory to the West and California.

In order to accommodate such travelers, there had to be taverns where one could obtain a night's lodging while awaiting passage on a boat. Connelly's Tavern on Ellicott Hill was one such place and King's Tavern was another. Lodgers at inns often had to sleep two or more to a bed.

CONNELLY'S TAVERN

KING'S TAVERN

Page 39

THE NATCHEZ TRACE

The Natchez Trace was the most important link between the Eastern United States and the lower Mississippi from Natchez to New Orleans. Horseback riders and foot travelers braved hostile Indians and highwaymen to pick up the trail at Nashville, Tennessee and come to Natchez to get aboard a boat headed for New Orleans. There were primitive taverns along the way, some run by half-Indians who provided bed and board. Mt. Locust Inn (right) above Natchez was one of these. There is also a craft shop (lower right) on the trace near Jackson where demonstrations are often given. Sometimes the Trace was just a foot path and often it was a narrow cut through the hills where robbers could hide out and jump unwary travelers. It was originally a trail between the Natchez, Choctaw, and Chickasaw Indian villages. President Jefferson ordered the Army to clear a frontier road.

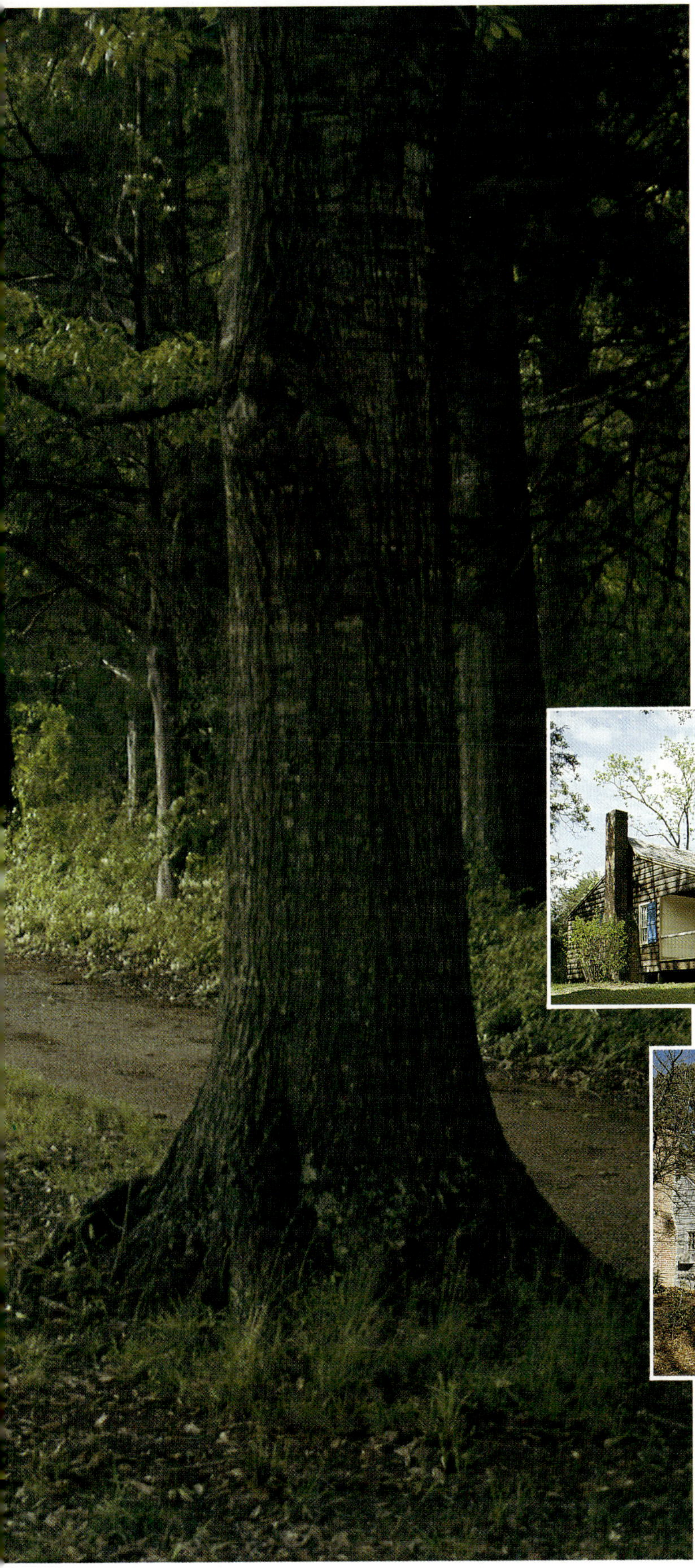
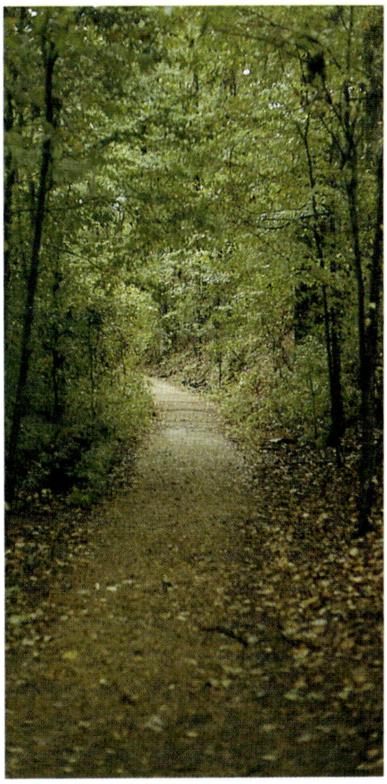
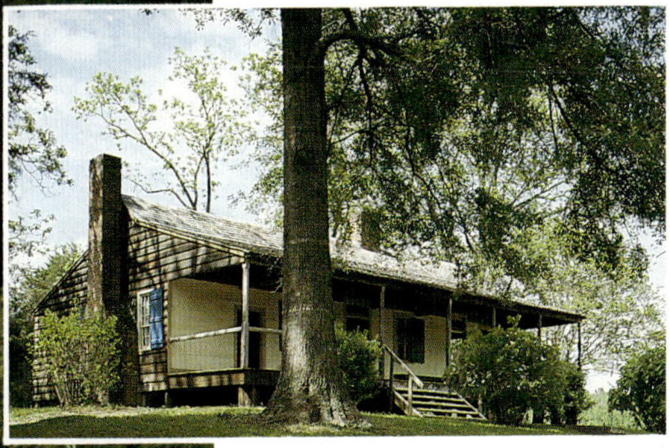
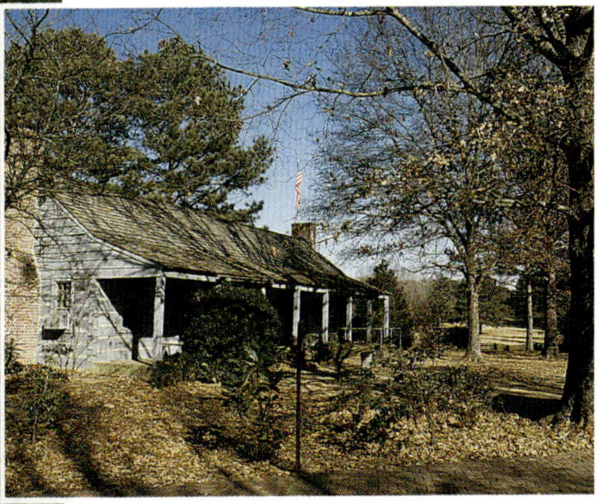

LONGWOOD

This spectacular octagonal architectural tour-de-force was begun by Dr. Haller Nutt before the Civil War. Many of his carpenters were Yankees. When war was declared, they laid down their tools and left. The house was never finished and for many years it was referred to as "Nutt's Folly", but has come into its own as a well-cared for and much appreciated landmark in Natchez.

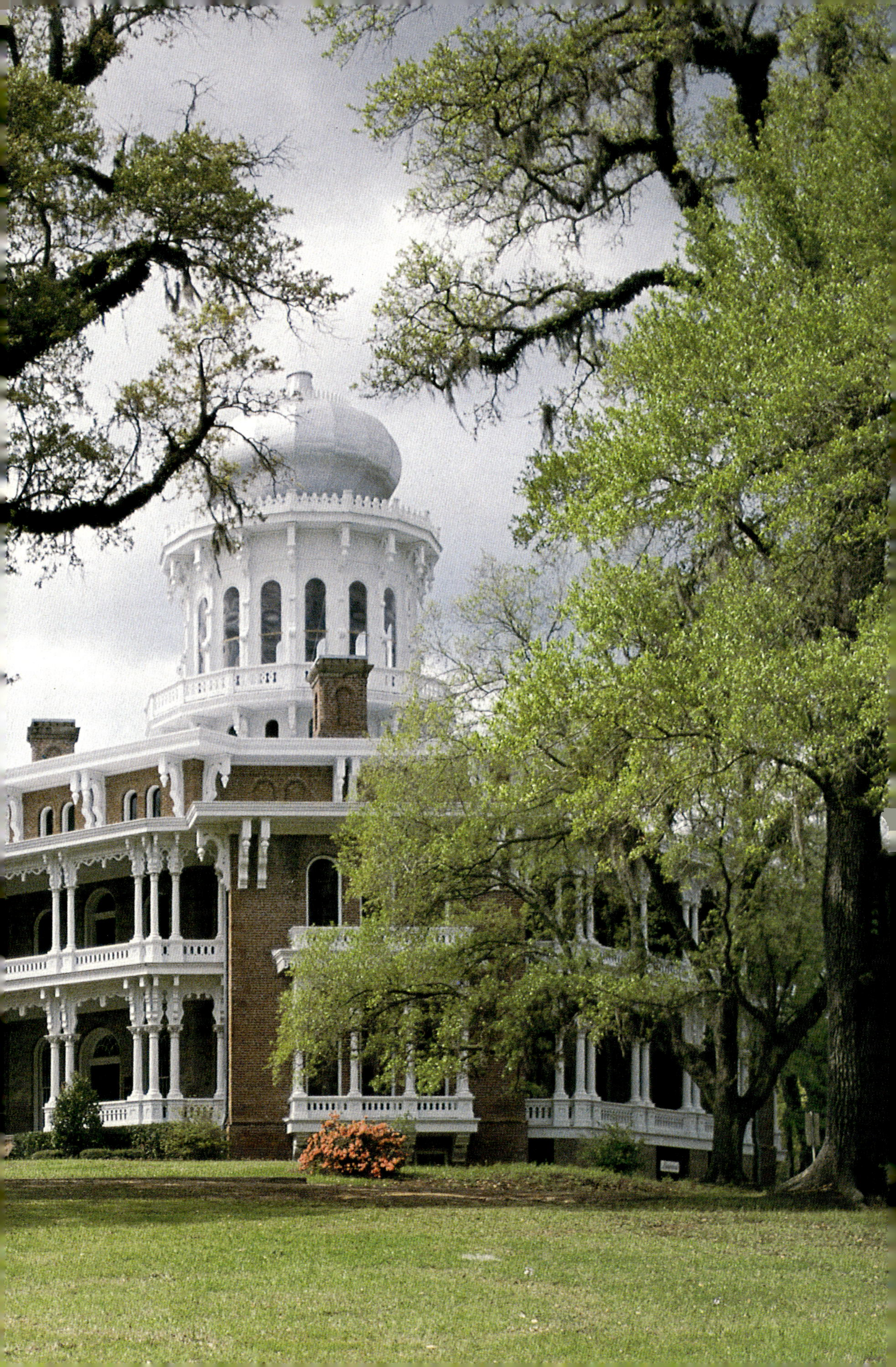

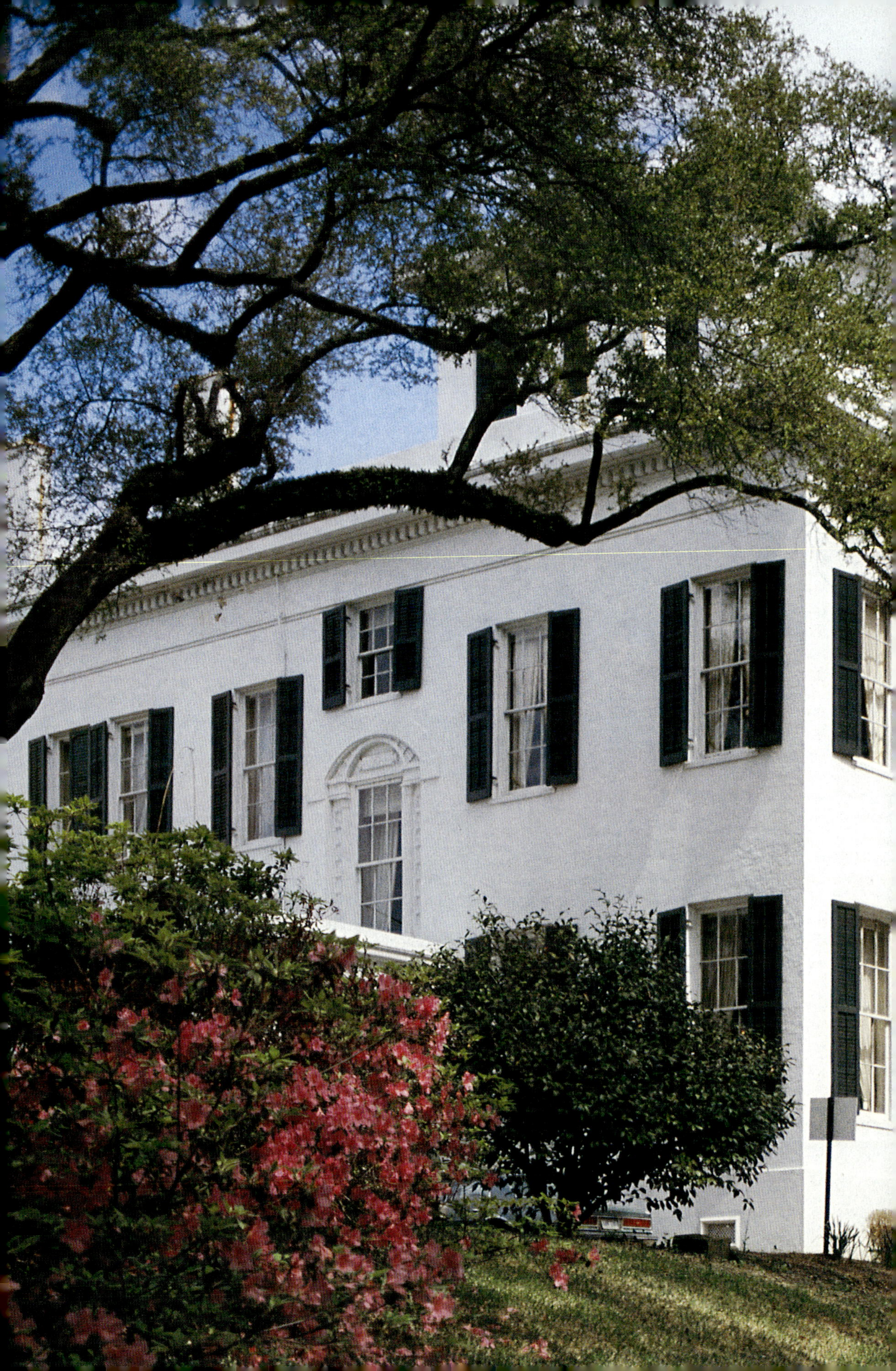

STANTON HALL

An Irishman built this outstanding example of 1857 craftsmanship. It is one of the grandest homes in Natchez, crowns a hill in the center of the city, and was whimsically referred to by its builder as an "ornament for the town". It has been the focal point for the Natchez Pilgrimage since this annual event began.

AUBURN

COLUMNS AND STAIRS

Auburn is an 1812 house. It is the centerpiece of beautiful Duncan Park and was a gift to the City of Natchez from the heirs of Dr. Stephen Duncan, the last owner. Duncan felt strongly about preserving the Union, and after the Civil War began, he went back to the North.

Operated as an historical preservation project by the Town and County Garden Club, the house is famed for its freestanding spiral staircase, with no visible supporting columns. As the sun sets in the west every evening, the pink brick takes on an ethereal glow and the four white columns are lighted as from within.

Dunleith is one of the largest antebellum homes in Natchez, and as pure a piece of classicism as one could find. It is truly a Greek Revival temple and has been the scene of many a festive event. When men came to Natchez to grow cotton, they became extremely wealthy in an extremely short time. Dunleith reflects that great wealth. Wide double galleries are symmetrical on all four sides of the house and each supporting column has its own separate base.

The interior is quite as elegant as the exterior. A two-story kitchen wing stands at the rear of the house, separate but attached, as was the tradition in Natchez. The immense lawn slopes down to a carriage house in the rear and a bayou at the side. One comes upon it suddenly, while driving into Natchez on a main thoroughfare, and the sight quite takes the breath away.

DUNLEITH

ROSALIE

TREASURIES OF HISTORY

Rosalie was built near the original Fort Rosalie which was the site of an Indian massacre. It is the property of the Mississippi Society of the Daughters of the American Revolution. Peter Little built the home in 1820 for his young wife, Eliza, who had become his ward when she was orphaned. General Grant spent three days at Rosalie near the close of the Civil War. At the side of Rosalie lies a peaceful green meadow which is the top surface of a river bluff standing above the old brick buildings in Natchez-Under-the-Hill. The meadow is surrounded by a split rail fence, and in the center stands a charming wrought iron gazebo.

D'EVEREUX

D'Evereux is a very pure interpretation of Greek Revival architecture. William St. John Elliott built the house in 1840, and some of his original furnishings, including a rare set of china, still remain in the home. A small square belvedere sits atop the peak of the roof.

MONMOUTH

LINDEN

Monmouth, 1818, was formerly the home of Mississippi Governor, General John A. Quitman, who was a hero of the Mexican War. This massive house has been a recent restoration, for it had fallen into disrepair, and the appointments now are splendid.

Linden, built in 1800, was once the home of Thomas B. Reed, the first elected U.S. Senator from Mississippi. The home has a two-storied central section, with a wing on either end, which permits a long gallery on the front.

Hope Farm, 1774-1789, once the home of a Spanish Governor, is surrounded by old-fashioned gardens. It was saved from destruction by Mrs. J. Balfour Miller and her husband. She was the founder of the annual Natchez Pilgrimage, and furnished her home with family heirlooms and rare Natchez antiques.

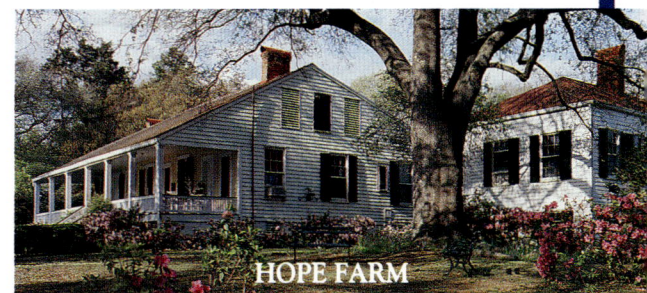

HOPE FARM

GRACEFUL HOMES FROM THE PAST

THE BURN

The Burn (left) is said to be the earliest Greek Revival house in the Natchez area. It was built in 1832 and takes its name from a Scottish word meaning brook, and just such a little brook used to trickle through the vast estate of the builder, John Walworth. It is a three-story mansion built on a hill, with one level below the main floor.

Evans-Bontura is the museum house of the National Society of Colonial Dames of America. It is much larger than the width of the front facade would suggest, and the upper and lower front galleries are totally framed in black iron lacework. The earliest section dates from Spanish times and was added to in 1830.

EVANS-BONTURA

NATCHEZ-UNDER THE HILL

The visitor to Natchez can leave Broadway, the street nearest the river on the top of the high bluff, and walk or drive down the very steep hill of Silver Street to the area called Natchez-Under-the-Hill. In early days there were four streets in this area, often visited by hard drinking, rowdy bullies from the flat boats. The river has taken three of them. The old brick buildings have been restored for restaurants, inns, and shops and the paddlewheel boats still deposit their passengers here.

SPRINGFIELD

LANDMARKS OF WARS

Just north of Natchez and adjacent to the Natchez Trace near the town of Fayette, Mississippi, lies Springfield Plantation. It is a part of the old Church Hill community. General Andrew Jackson and Rachel Robards are said to have been married at this restored 18th Century house. Around the bend is an ancient church (right), high on a hill, that looks as if it came out of an English story book.

Farther up the main highway, at the town of Lorman, is the old Lorman Country Store, which has all of its original furnishings, including a tall, round iron "shoe tree" that turns on a windup clockwork to display the shoes.

On a side road out from Port Gibson, one arrives at Grand Gulf, which once ranked third in the state as a commercial port. It was bombarded at the same time Vicksburg fell to U.S. Grant. All that remains today is Grand Gulf Military Park which includes the remains of Civil War battle sites and a museum containing prehistoric and military artifacts.

OLD LORMAN COUNTRY STO

GRAND GULF MILITARY PARK

WINDSOR RUINS

Port Gibson, Mississippi was the town that General Ulysses S. Grant proclaimed as "too beautiful to burn" as he marched south. Both sides of the main street are lined with elegant homes. One fine example of Greek Revival architecture is Oak Square, graced by six fluted Corinthian columns.

The unusual First Presbyterian Church has a hand atop the steeple, pointing the way to Heaven.

Just south at Lorman, is another handsome columned house, Rosswood. The largest plantation house ever built in the area was Windsor that was destroyed by fire shortly after it was built, but visitors still go to gaze at the ruins.

OAK SQUARE

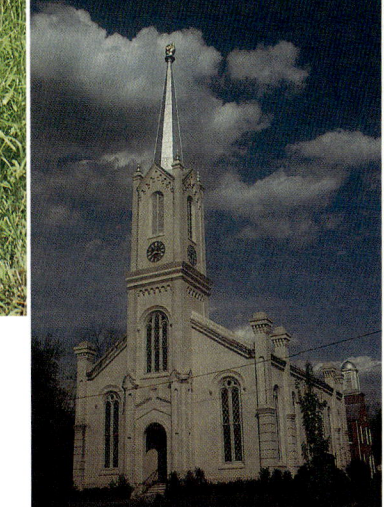

ROSSWOOD

TWIN BRIDGES

Although the two bridges are not identical twins, both were needed to carry the flow of traffic from Vicksburg, Mississippi to the Louisiana side of the great river. An Excursion Boat, the Jefferson Davis, brings visitors out on the Mississippi from Vicksburg.

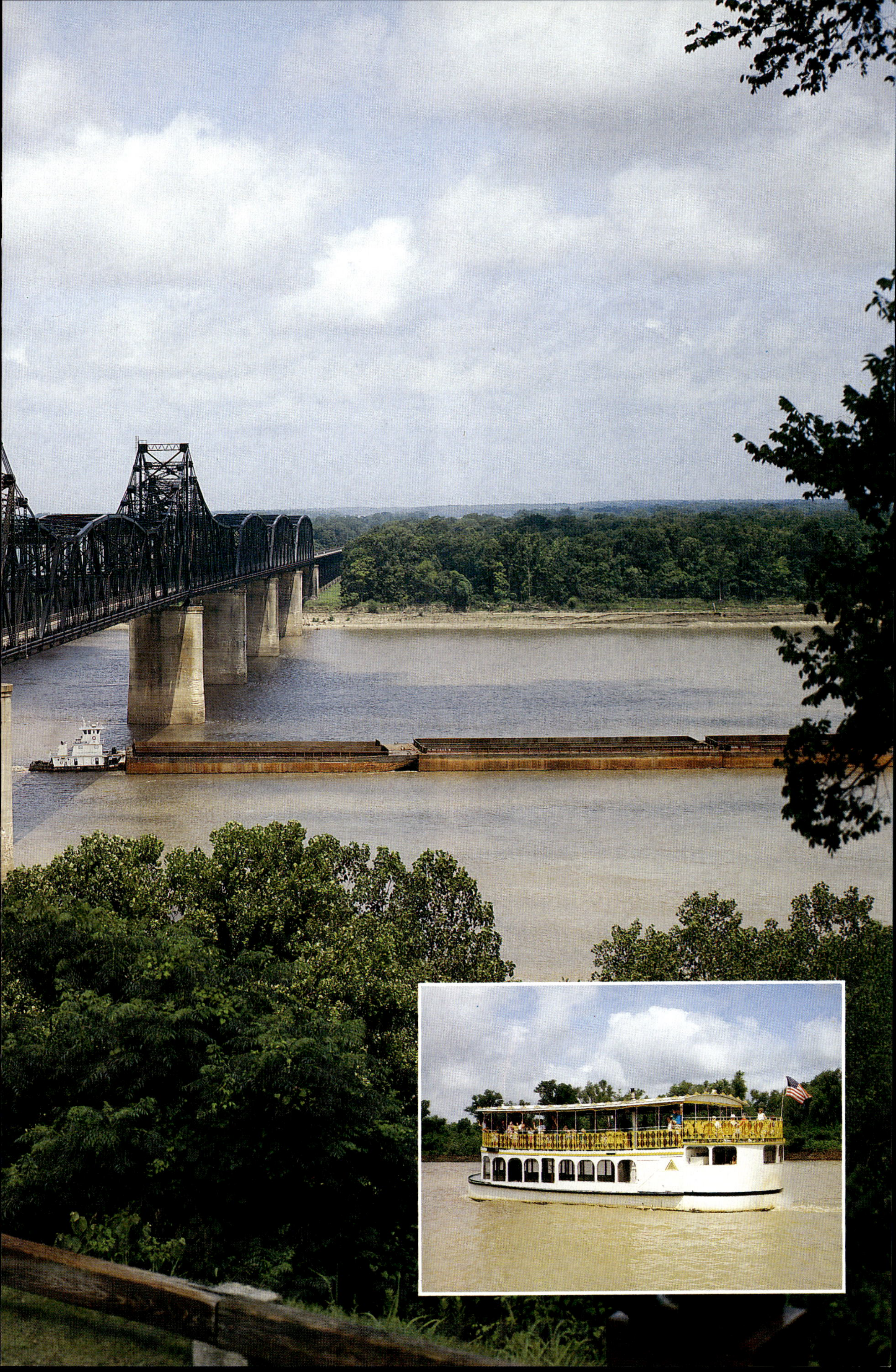

THE RIVER AND GENERAL GRANT

In every war that was fought in America, from the time of the Revolution, each side believed it was imperative to control the Mississippi River. Spain, being in control of the Baton Rouge area in 1776, sided the colonists against the British. Andrew Jackson became famous in the War of 1812 when he drove off the British at the Battle of New Orleans.

Near the end of the Civil War, it was critical for Union troops to reopen the river to commerce, but they could not conquer the battery of cannons on the 200-foot bluffs at Vicksburg. After choking the city off from all aid, Grant kept the town in a state of siege for a month and a half, until illness, wounds and lack of provisions forced it to surrender. The townspeople had retreated to caves in the bluffs during the shelling, stifling in the humid summer heat. Grant even went to the extreme measure of digging an enormous canal so that he could get closer to the city. It is still known as "Grant's Canal."

Arms were laid down on July 4, 1863 and an impasse had been reached at which the South could not hold out much longer. Vicksburg, and Port Gibson to the south, were both major ports and both fell at the same time. The back of the Southern effort had been broken. This event was truly a turning point in history.

There is now some danger that "Old Man River" could, at some future whim, take another new path to the Gulf, leaving many port cities high and dry. The U.S.

EMMA BALFOUR'S DIARY

Army Corps of Engineers maintains working scale models of different sections of the river at the Waterways Experiment Station at Vicksburg. They have pumped grout in to harden and to strengthen the danger point in the Old River Control Structure in order to hold this mighty river on its present course.

At the time of the Civil War, Balfour House was the home of a courageous lady named Emma Balfour. She was the wife of Dr. William T. Balfour who had built the house in 1835. Emma kept a diary which chronicled the terrible Siege of Vicksburg. The diary remains as a vivid description of a genteel and courageous lady who lived through the day-to-day strain of the bombardment, and gave shelter to many. She tells of staying in the hot, humid caves for so many days that she could no longer bear it, and vowed to remain in her home, even if she were killed. She tells of the men of the town having to kill mules and livestock because they had nothing to feed them.

After the barrage of artillery fire, Grant followed with a massive infantry assault, which was again beaten back by the southern General John C. Pemberton. Grant then laid the siege which lasted for 47 days. Pemberton learned that Grant was planning another massive assault and concluded that surrender was inevitable.

In spite of the repeated heavy shelling it received, with cannon balls actually entering the home, Balfour House withstood the ravages of war, as did many other fine homes in Vicksburg. Today, gracious hospitality is still a part of this community, and friendly residents welcome traveling "Yankees".

There is no way to estimate the importance of the Mississippi River to this land. Surely no history book could be written if it omitted the effect that "Old Man River" has had on the United States of America.

BALFOUR HOUSE

CEDAR GROVE

THE GRAND HOUSES OF VICKSBURG

Cedar Grove sits on one of the many "terraces" that climb the hill from the river at Vicksburg. Because of this elevation, it presents a monumental image to the visitor looking up at it. The house is flanked by two quite lovely terra cotta pieces of statuary, and the wrought iron embellishments add to its beauty. The house was built in the 1840s by John A. Klein. There are double galleries on the stately home, which are supported by four enormous columns. There is an interesting little bay window in the wing to the left, which makes for a charming room inside that wing.

General Grant once slept here, in spite of the fact that his siege almost destroyed the place. There is a cannon ball still imbedded in the wall of the parlor, and a large hole remains in the floor, covered with glass, so that visitors can view where a cannon ball went through the floor into the basement.

Grey Oaks is actually a newly-reconstructed home, but it is nevertheless an "old" home. It was an antebellum home in Port Gibson, but it was totally dismantled and gently moved to Vicksburg, where it was rebuilt in 1940 on the bluffs along the river, just as all old Vicksburg homes were. This eight-columned house with its many tall shuttered windows is a fitting companion to the splendid houses of Vicksburg. Gracious gardens and vast green lawns surround the house, and the companionship of old trees make it seem as if it has been there forever.

GREY OAKS

McRAVEN

The delicate columns of McRaven and its classical facade mask a home that was built in three parts during three distinct periods in time. It is one of the most interesting homes one could possibly visit, for it is like a trip going back in time.

The back section was an 1825 frontier cottage built on the Newitt Vick plantation when Vicksburg was founded, and its interior has been preserved in a fairly primitive state. The mid-section is early pure Greek Revival and was added in 1836 by Sheriff Stephen Howard.

In 1849, a new owner, John Bobb, added a Philadelphia Townhouse as the third section in "high" Greek Revival style, complete with elegant gas chandeliers. Interior steps lead up to a slightly higher level from the mid to the front section. To continue the theme of "High Greek", there is even a Greek Revival privy in the backyard.

The Duff Green House is most notable for its abundance of lacy wrought iron columns and railings which both support and frame the double galleries at the front. Underneath the iron lacework is a house that is Greek Revival in character.

It was built in 1840 and is known to have been used as a Confederate hospital during the siege of Vicksburg. Mrs. Green, who was the mistress of the home, apparently thinking there would be no privacy in a hospital full of wounded men, took refuge in a nearby cave during the bombardment and gave birth to a child there. She bestowed the appropriate name of Siege Green upon him.

DUFF GREEN HOUSE

Page 59

MONUMENTS TO MEN OF VALOR

The Vicksburg National Military Park is dedicated to the memory of the courageous soldiers of both the Union and the Confederate armies who fought in this tragic battle. It is one of the best preserved and maintained military memorials in America and is a unit of the National Park System. The park has been laid out over the crest of the bluffs, in the actual location where much of this tragic battle was fought. Men charged over these hills, and hid in the valleys, dug trenches and placed cannons; and to visit it is like standing with ghosts of the past. Cannons still stand in rows where some of the original batteries were placed. Wonderful obelisks of marble, statues of bronze, and white temples have been placed in the park by the various states of the North and South that were involved in the Civil War. General Ulysses S. Grant on his horse, and General Robert E. Lee standing proud, are both cast in bronze. There are small marble stones scattered throughout the hills that commemorate specific battalions, and seemingly thousands of small white headstones mark the graves of fallen soldiers on a green expanse of lawn that spreads out below the bluffs.

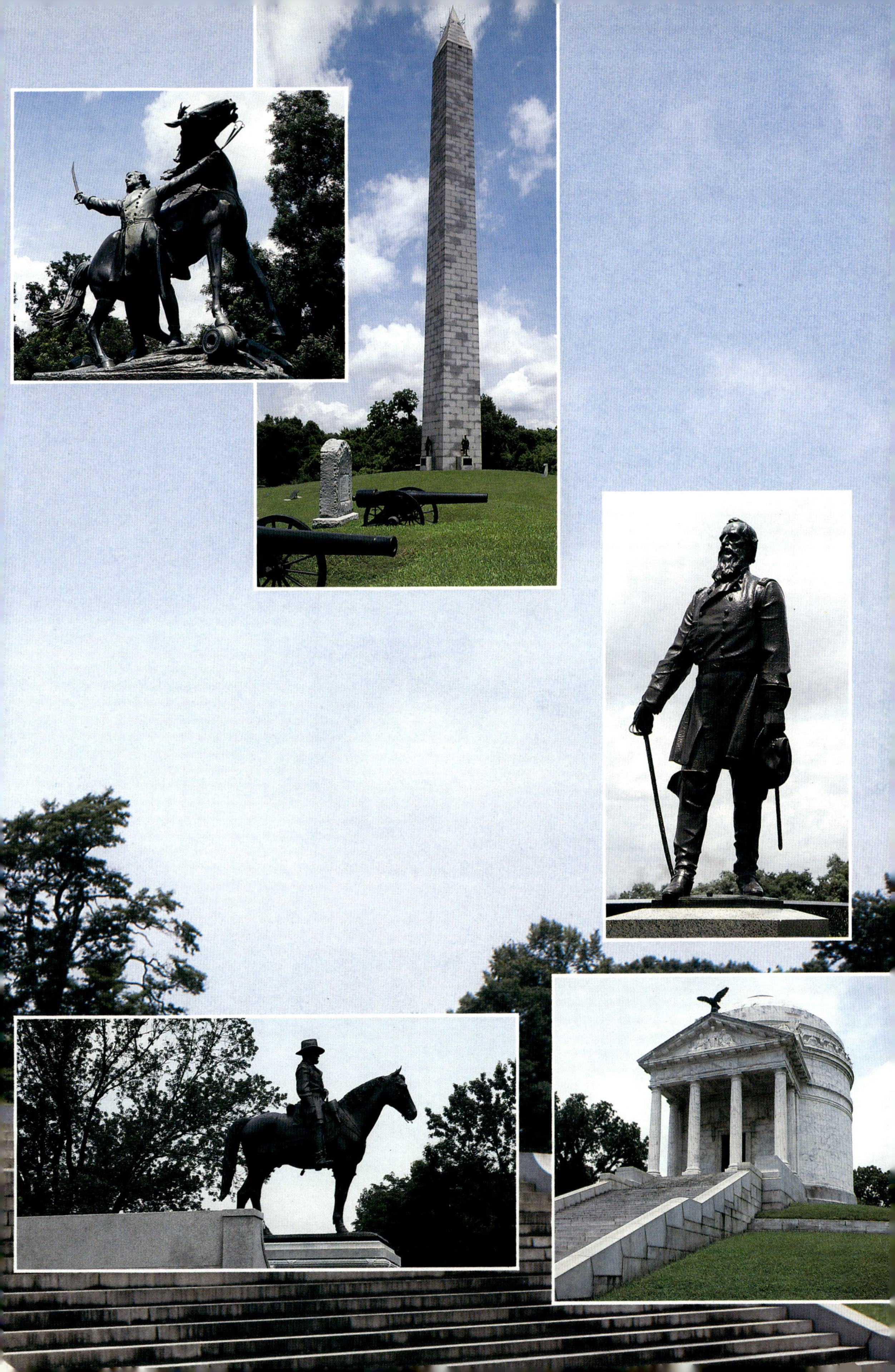

AT THE TOP OF THE DELTA

Crossing the bridge from Vicksburg into Louisiana and traveling northward toward Arkansas, one finds Lake Providence, the oldest town north of Natchitoches in Louisiana, on just one more of the river's oxbow lakes. The name comes from the bent piece of wood that passes under the neck of a draft ox, and the shape happens when a loop of the river is cut off. There are many others in this area, because when the river reached the lowlands of Louisiana and Mississippi, it meandered a lot and created large loops. Lake Providence is a restful place where many people have camps and fishing piers among the cypress trees. The black soil of the Delta continues this far north.

Dropping back southward to Epps, one finds the Poverty Point State Commemorative Area. This is the site of mounds of the earliest culture yet discovered in the Mississippi Valley, dated between 700 and 1700 B.C.

Farther south, near Newellton, a very early house has been preserved in the Winter Quarters State Commemorative Area. This nineteen room white frame plantation house was used as headquarters by General Grant during the siege of Vicksburg. It was originally the home of Dr. Haller Nutt who later built the magnificent octagonal Longwood across the river in Natchez, Mississippi.

Page 62

FLOREWOOD RIVER PLANTATION

Back on the Mississippi side of the great river, one is approaching Memphis, and this is still Delta country. Greenville is a typical delta town. It is worth sidetracking a short distance east to Greenwood to see Florewood River Plantation State Park, which is a reconstruction of a typical river plantation, complete with outbuildings and a small swamp with cypress trees showing their knobby knees. Back at Greenville, visitors can stop at the Welcome Center in a facsimile of a river boat. Other state parks here are the Great River Road State Park (below), Winterville Mounds Historic Site (Indian) near Greenville, and Leroy Percy State Park at Hollandale.

RIVER SWAMP

RIVER ROAD STATE PARK

ORDER FORM

If you would like to order additional copies of this book or sample some of our other fine products, please fill out the form below and mail to: YOUR POINT OF PURCHASE RETAILER
OR
R.A.L. ENTERPRISES
Suite 136, 5000 A West Esplande Ave.
Metaire, LA 70006

TITLE		COST	QUANTITY	TOTAL
PLANTATION COUNTRY GUIDE	64 pgs.	5.95		
Cajun Country Cooking	64 pgs.	5.95		
Favorite Recipes from New Orleans	64 pgs.	5.95		
Southern Seafood Sampler	64 pgs.	5.95		
Favorite Drinks of New Orleans	32 pgs.	3.95		
Cookin' on the Mississippi	64 pgs.	5.95		
New Orleans - Birthplace of Jazz	56 pgs.	6.95		
New Orleans - Crescent City	32 pgs.	3.95		
Laminated New Orleans Placemats	Set of 4	7.95		
Laminated Louisiana Plantation Placemats	Set of 4	7.95		
Laminated Mississippi Plantation Placemats	Set of 4	7.95		
New Orleans Coloring Book	32 pgs.	3.95		
Louisiana/Mississippi Coloring Book	32 pgs.	3.95		
Recipe Box Cards	Set of 10	5.95		
			Postage & Handling	1.50
			Total	

☐ Check Enclosed   ☐ Visa   ☐ Master Charge   ☐ American Express   ☐ Discover

Card Number _____ Expiration Date _____
Name _____ Address _____
City _____ State _____ Zip _____ Daytime Phone (_____) _____

All items are satisfaction guaranteed and your purchase price will be promptly refunded if returned within 30 days.
Please allow two-four weeks for delivery. No foreign orders please.

ORDER FORM

If you would like to order additional copies of this book or sample some of our other fine products, please fill out the form below and mail to: YOUR POINT OF PURCHASE RETAILER
OR
R.A.L. ENTERPRISES
Suite 136, 5000 A West Esplande Ave.
Metaire, LA 70006

TITLE		COST	QUANTITY	TOTAL
PLANTATION COUNTRY GUIDE	64 pgs.	5.95		
Cajun Country Cooking	64 pgs.	5.95		
Favorite Recipes from New Orleans	64 pgs.	5.95		
Southern Seafood Sampler	64 pgs.	5.95		
Favorite Drinks of New Orleans	32 pgs.	3.95		
Cookin' on the Mississippi	64 pgs.	5.95		
New Orleans - Birthplace of Jazz	56 pgs.	6.95		
New Orleans - Crescent City	32 pgs.	3.95		
Laminated New Orleans Placemats	Set of 4	7.95		
Laminated Louisiana Plantation Placemats	Set of 4	7.95		
Laminated Mississippi Plantation Placemats	Set of 4	7.95		
New Orleans Coloring Book	32 pgs.	3.95		
Louisiana/Mississippi Coloring Book	32 pgs.	3.95		
Recipe Box Cards	Set of 10	5.95		
			Postage & Handling	1.50
			Total	

☐ Check Enclosed   ☐ Visa   ☐ Master Charge   ☐ American Express   ☐ Discover

Card Number _____ Expiration Date _____
Name _____ Address _____
City _____ State _____ Zip _____ Daytime Phone (_____) _____

All items are satisfaction guaranteed and your purchase price will be promptly refunded if returned within 30 days.
Please allow two-four weeks for delivery. No foreign orders please.